LEO
the maker prince

JOURNEYS IN 3D PRINTING

by Carla Diana

This book is dedicated to my father, Joseph C. Diana,
who taught me to find wonder in everything.

LEO the Maker Prince: Journeys in 3D Printing
by Carla Diana

Visit **LeoTheMakerPrince.com** to learn more, or go to **Thingiverse.com/LeoTheMakerPrince** to download the objects in this book.

Printed in Canada.

Published by Maker Media, Inc., 1005 Gravenstein Highway North, Sebastopol, CA 95472.

Maker Media books may be purchased for educational, business, or sales promotional use. Online editions are also available for most titles (http://my.safaribooksonline.com). For more information, contact O'Reilly Media's corporate/institutional sales department: 800-998-9938 or corporate@oreilly.com.

Editor: Brian Jepson
Story, illustration, and object design: Carla Diana
Book design: Nicholas Lim
Story editor: Cindy D. Hanson
Photography: Claudia Christen
Design and production assistance: Alexa Forney

December 2013: First Edition

The Make logo and Maker Media logo are registered trademarks of Maker Media, Inc.

ISBN: 978-1-4571-8314-0

[TI]

PREFACE

2013 has often been called "the year of the 3D printer." While the technology for automatically fabricating solid, 3Dimensional objects based on stereolithography has been around for decades, it was so expensive that it could only be used in industrial and commercial applications. Only very sophisticated manufacturing facilities and design firms could afford one, and the parts themselves were expensive to produce, compared to the price of a similar, mass-produced part. They were also delicate and highly limited in terms of material qualities.

What led up to this change in 2013 was the fact that the DIY movement encouraged people tinkering with making their own kinds of electronic, customized devices. Built from off-the-shelf components, housed in laser-cut panels, and assembled by hand, these devices formed the foundation for visions of new products. One such device was the MakerBot, launched in 2009 by MakerBot founders Bre Pettis, Adam Mayer, and Zach "Hoeken" Smith, who created the first affordable 3D printer at a hacker space in Brooklyn, N.Y., NYC Resistor. With their vision of creating "3D printing for the masses," the MakerBot marked what might be called a "Macintosh moment," referring to the turning point when a vision of turning an expensive piece of equipment made for business use only (the computer) into a central and ubiquitous part of everyday life became reality.

In 2009 the first MakerBots were sold at a price of $1300, over one tenth the price of the Dimension uPrint, one of the leading professional 3D printers at the time. By mid-2013, 30,000 MakerBots were out in the world, and more than a dozen other companies creating different types of inexpensive 3D printers had formed.

While the vision of 3D printers for the home has become a reality, there are still many predictions for how exactly they will be used or what impact they will have on everyday life. Some theories say people will download their products, making a single object from a digital file rather than going to a store and buying a mass-produced item off a shelf. Others predict the printer will be used to enhance or modify existing objects, allowing people to customize products to suit their individual tastes and needs. And others believe that the 3D printer will spawn new types of "microfacturing" businesses where entrepreneurs with new ideas can quickly build, replicate, and distribute new products themselves, with very little overhead and startup cost.

This book is a celebration of those possible futures. Though the visions may be a little blurry today, clouded by the limitations of current machines and material combinations, they are exciting glimpses into how putting new technologies into the hands of regular people like you and me can change the world.

In sharing these visions through a book intended for 5 to 8 year-olds (and creators of all ages), we hope to not only mark this moment in time, but to inspire a rich future of invention for decades to come. By putting these ideas into the hands of children, we can jumpstart our ambassadors for the future.

With excitement and a loving nod to Antoine de Saint-Exupéry, we invite you to enjoy *LEO the Maker Prince*.

When I was small,
  I used clay to create...

I made a spiral slide for a bee. The grownups said, "That's very nice, Carla. You made a pipe."

So I tried again, creating version two, and the grownups said, "That's very nice, Carla. You made a bolt."

But I was determined. Version three included seesaws for caterpillars, jungle gyms for ants, and sleeping chambers for butterflies. And again, I was misunderstood. I wondered, "How can they not see what I see?"

And so, right then and there, at the age of eight, I gave up what might have been a magnificent career as a sculptor. I put away my clay, packed up my masterpieces, and proceeded to move on to more practical pursuits.

In the end, I became an accountant. Instead of talking to people about seesaws for caterpillars, I talk about dollars and sense, celebrating the importance of balance sheets, annual reports, and the ever important ROI, return on investment.

I moved to New York City (the capital of all things math and money) with all of my boxes (including my misunderstood sculptural masterpieces) and took a job as a bean counter, number cruncher and money mover at the firm of Johnson, Smith and McElroy. And I never thought about sculpture or art again, until...

That day of the big storm,

HURRICANE SANDY

● HOME

You see, I am an accountant who loves to ride my bike everywhere, even if it isn't the most practical means of transportation. Biking allows me time to turn off all the numbers and ponder, ever so slightly and lightly, the worlds of butterflies and bees. That is how I found myself biking back from a client meeting that had taken me from Manhattan across the Brooklyn Bridge and into the heart of New York City's second largest borough, Brooklyn, or, the other Manhattan. By the time the meeting was over, Hurricane Sandy was percolating and proving punctual, and I was pedaling as fast as possible to get back home safely.

When all of a sudden, a gust of wind took me, and the next thing I knew, I was splat on the pavement. The world was spinning, and blurry. I believe I blacked out.

Because when I came to I could just about make out a silhouette looming over me. Too round to be a person and looking a little like a—no, it couldn't be. Or could it?—a robot? Here? Now? I thought to myself, "How hard had I actually fallen?"

ME

MEETING
SITE

But it was a robot! I admit to being
a tiny bit frightened, but there was
something very intriguing about
this little mechanical character. He
had chunky legs, and straight arms
that gripped a small tray in front of
him. Behind him was a special kind
of backpack that held spools of what
looked like black and white wire. A
miniature version of the mechanical
arms you see on construction
equipment emerged from behind him
like a swirling and swinging tail.

And then the robot spoke. "Hello there! My name is Leonardo, or LEO, for short. What's yours?" I figured it couldn't hurt to respond.

"I'm Carla."

"Hello, Carla! It's a pleasure to make your acquaintance. Looks as if we are in the midst of a fast approaching storm, so would you please draw me a sheep?"

"Excuse me?" I was mystified by such an artistic request at such an inappropriate time, but I didn't want to be impolite. "I'm sorry, but I believe you must have mistaken me for someone else."

"Oh no." he said, "I don't actually make mistakes. You are just the person I'm looking for," he insisted. "The coordinates indicate that this particular superstorm will be bearing down on us rather quickly, so perhaps we should get right to it. Would you please draw me a sheep?"

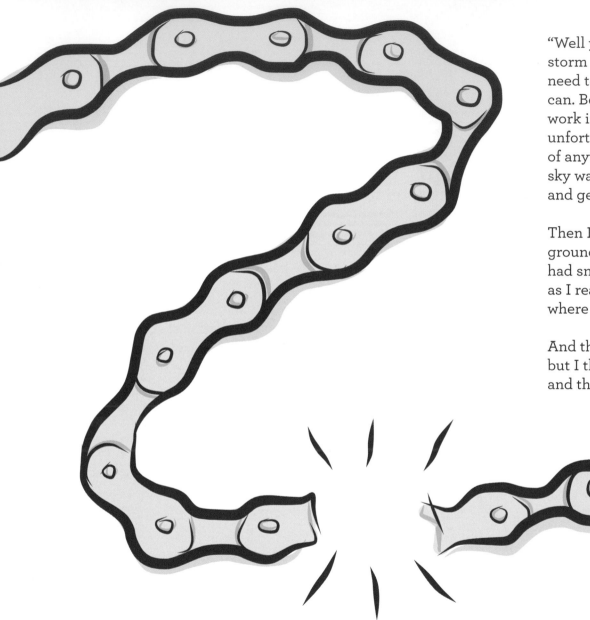

"Well you're not kidding about the storm and that's why we both really need to get home as soon as we can. Besides, I'm an accountant. I work in finance. Which means that, unfortunately I don't draw pictures of anything, let alone a sheep." The sky was filling with thunder clouds and getting darker and darker.

Then I pulled my bike up off the ground and noticed that the chain had snapped. Panic ran through me as I realized I was a long way from where I needed to be.

And then, I'm not at all proud of it, but I threw a bit of a fit. Right then and there in the street.

"Oh my," LEO said, "I haven't been programmed to process such behaviors!"

"Sorry. I just don't know what I'm going to do."

"Well, I feel absolutely certain of your ability to draw me a sheep. So why don't you do that? Please?"

He was so persistent. And I did need to get a hold of myself. Maybe this robot distraction would be a good way to calm down for a moment so I could think clearly again and figure out how to get home.

"Okay, LEO. Fine. You drive a hard bargain", I said, "I'll draw you a sheep. Here goes." I quickly pulled some paper and a ball point pen out of my knapsack and began drawing the best sheep I could muster, given my accounting education. It felt so foreign at first, but then something from my sculpting days began to take hold. My imagination took over. Lines poured out through the pen and onto the paper, and lo and behold, I had drawn a sheep!

"PERFECTO!"

LEO exclaimed, and extended the tray he was holding so that it lay flat in front of me. It seemed to be an invitation of some sort.

"Here?" I asked. "Precisely!" he said. So I put my drawing on top of the tray, wondering what this little robot could be up to. He tilted it back towards him just slightly as projected light ran back and forth across the page. It was kind of spectacular.

"Scanning complete!" he said with authority, and tilted the drawing so I could take it back.

LEO then lifted his arms and the tray he was holding became a horizontal platform above his head. The tail-like contraption attached to his back began to move and positioned itself in the middle of the platform. I was mesmerized. Magic was unfolding right in front of me.

"Heating up the filament now," LEO declared, and I began seeing waves of heat radiating from the nozzle. "LEO? Hello? Oh dear. Is that, um, supposed to happen?"

"Part of the process! Rest assured! Not to worry." And a few moments later he proclaimed, "Beginning build!" His announcement was filled with such excitement that I swear he sang a little tune.

Surprisingly, the nozzle above his head began to move in an elegant dance as melted plastic came out of it and landed on the platform. The spool on his back turned ever so slightly as the wire was drawn into his tail, emerging moments later as a thin line of hot, melted plastic.

LEO was generating amazing heat and precise movements. But to what end? I had no idea. And then, four pools of plastic appeared on the tray. "Sheep feet?!" I exclaimed, as I realized that the black squares matched the bottom view of the sheep I had drawn.

More graceful robot arm ballet happened before my eyes, as layer after layer of plastic was placed on top of one another.

Eventually I could see the sheep's legs take form, and then the body, and finally the head!

"AND VOILÀ"

LEO exclaimed, as he presented me with a fully formed sheep.

"That was amazing!" I had almost forgotten my aches and scrapes and the fact that I was stranded. And even though I could still see the broken chain on my bicycle out of the corner of my eye, it was the last thing on my mind.

I needed to know more. "How did you do that? Where did you come from?"

"I come from the land of Kings," he said.

"Kings?"

"Well, some call it 'Kings County', and most call it 'Brooklyn', but I prefer the royal aspect. Even though my name is LEO, I'm sometimes called the Maker Prints. Get it?"

"Uh, no, not really. 'Prince?' As in soon to be king?"

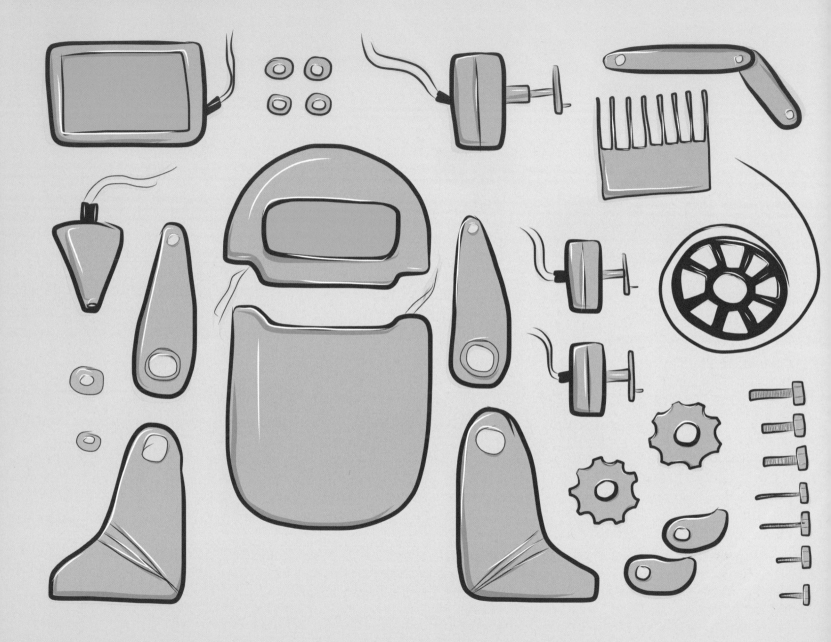

"No, I am a 3D printer. I make 3D prints—that's 'ts'—from files that are put into my memory by following the code in the file and moving my tail and tray to and fro while melting plastic through my nozzle. The melted plastic lands on my tray in layers, one at a time, until eventually a physical form is built up vertically. I was constructed from spare parts and electronics in a warehouse space in Fort Greene, and was on my way to an artist's home." LEO pointed to an address label attached to his back. I read it aloud, "Sally Harris, 166 Seventh Street, Brooklyn, NY 11215."

"Exactly. But my delivery truck must have hit a particularly bad pothole because I felt a big bump and then fell off the truck and onto the street." I stopped him with a tad bit of skepticism. "Wait. You *fell off a truck*?"

"Oh yes," Leo confirmed, "The doors flew open and I was hanging on for dear life, but I went toppling out instead. Sadly, the driver was unaware and just kept going. I'm hoping he'll realize his error soon and return for me. But in the meantime it's absolutely vital that—"

"Hold on," I stopped him, in an attempt to try to wrap my head around all of this new information. "You, the heated nozzle, the sheep, I mean, wow, I've never seen anything like it! Are you a magical, one-of-a-kind creation?"

"Oh, goodness me, no. Hardly!" he scoffed, looking as modest as a robot could. "There are actually others like me, and not exactly like me, but with similar intents and processes, being made every day. We're getting shipped all over the world...and, of course, Brooklyn."

"And you're all making...tiny sheep?"

"Tiny sheep? All of us? Of course not. No, no, no." he chuckled. "Sally Harris is a sculptor who makes magic gardens. And every magic garden needs a sheep. So I decided to have you draw one that I could put in my memory. And what a lovely sheep it is, indeed! A sheep, you see, can always find his way home. I'm simply thinking ahead because the storm is coming and in case the driver doesn't return."

"Oh, I see." And I thought to myself, "I like the way this robot thinks."

"But to answer your question, artist sculptures are just one thing people are doing with robots like me."

"Really?"

"Unequivocally! We are all quite flexible in nature and industry, and when a good human comes by and gives us a drawing of an object, we get our gears ready, heat up our nozzles, and position our trays to prepare to make it into a real thing."

"I had no idea!" I said. And I really didn't. "A drawing into a real thing? I've never heard of anything like this."

"I'm the kind of robot with a plastic filament on my back and a heated nozzle for a tail. But I have friends in other neighborhoods, doing all sorts of 3D printing work in different sizes and with different materials." I can show you a few examples." And with that, he flipped his tray over and showed me a map of Brooklyn, with one block in particular, on the side, highlighted.

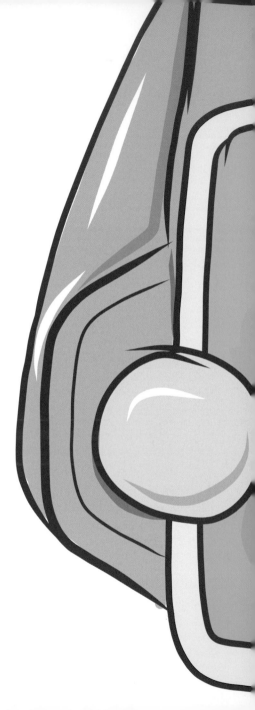

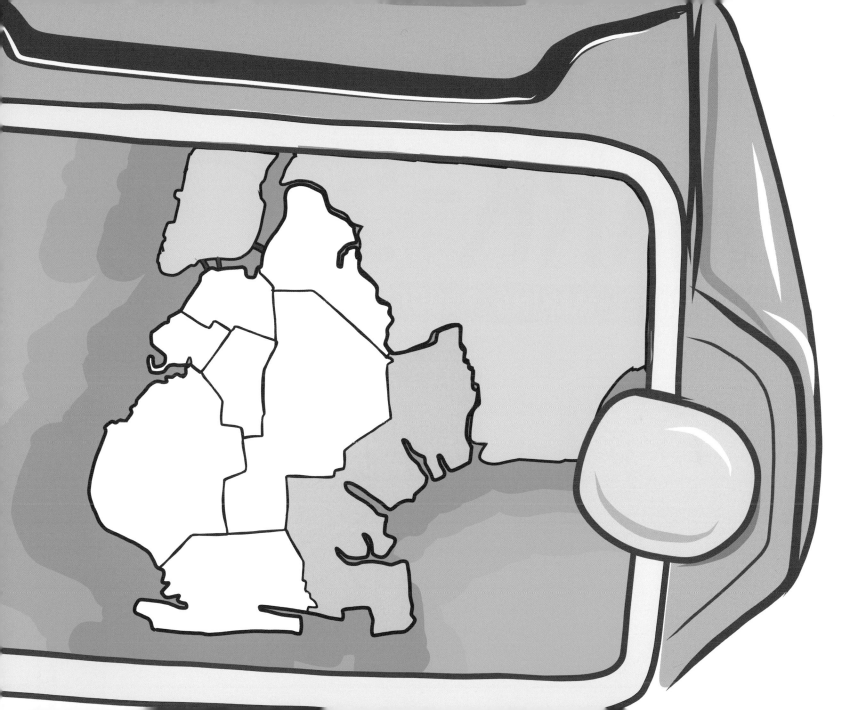

BRENDAN &
MARK-5
THE MODEL-MAKING ROBOT

MARK-5 [märk fīv]
Process: FDM*
Description:
Classic 3D printer
Features:
High-resolution output

* FDM, or Fused Deposition Modeling uses thin streams of
melted plastic filament to gradually build up a physical form.

BRENDAN

Age: 60
Location: East Brooklyn
Likes: Spaghetti nights with his family
Talent: Knows twelve computer programming languages

In the northeast of Brooklyn, my friend Mark-5 works with Brendan. Brendan is an inventor, and has been his whole life! Mark-5 is one of my oldest and best friends. He helps Brendan create the first versions of his inventions that will eventually be produced in factories. These models are sometimes called 'prototypes' and are used to help test how the product might work before making the final design."

"Prototypes?"

"Prototypes."

"Sounds efficient."

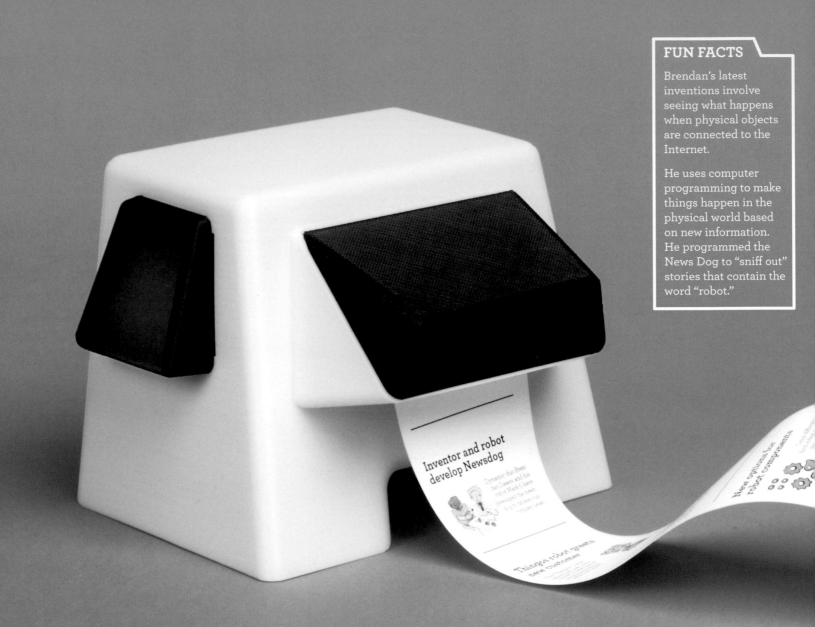

Inventor and robot
develop Newsdog

Dynamic duo Bren-
dan Dawes and his
robot Mark 5 have
developed the news
dog to deliver out
robot news.

Things robot greets
new customers

"Immensely so. Last week, they were working together on a model for a robot dog that will print a special version of current events. They call it 'News Dog'. News Dog can be programmed to give news on a specific topic. For example, robots. We've been in the headlines quite a bit lately."

"Seriously?"

"Absolutely! We robots can collect information about new trends as well as make things that are practical and useful. We also make items that are beautiful and that adorn desktops as well as people."

Heated nozzles and
3D printing robots

Ambassador greets
robot diplomat

Robots to build on
moon surface

Inventor and robot
develop Newsdog

In the middle of Brooklyn, near the big park, is H1-H0, a robot I've known for years. He works with Stephanie, a jewelry maker. She makes her designs on a computer and then H1-H0 prints them out in gold, silver and bronze. Before H1-H0, Stephanie depended on a outside factory to make her earrings and necklaces. Now she can make them herself and ship directly to her customers. She keeps H1-H0 busy morning, noon and night making hundreds of copies of her jewelry pieces.

"Question!" I stopped LEO for clarification. "3D printers can work with different kinds of metal?"

"Yes," LEO assured me. "We are quite versatile in our work methods and materials."

STEPHANIE
Age: 35
Location: Park Slope
Likes: Bringing a sketchbook wherever she goes
Talent: Expert in biology and math

"And since Stephanie is infinitely inventive, she loves working with H1-H0 to experiment. And he is always happy to comply. She especially likes to use mathematics to create different shapes. Then H1-H0 builds them using the formulas Stephanie inputs."

"Formulas?! As in mathematical formulas? Stephanie really uses math?"

"Of course." Math is something we robots handle extremely well."

"I'm good at math!"

"I'm sure you're good at many things." Despite the predicament I was in, LEO was making my day.

"And are there more kinds of robots?"

"Of course!"

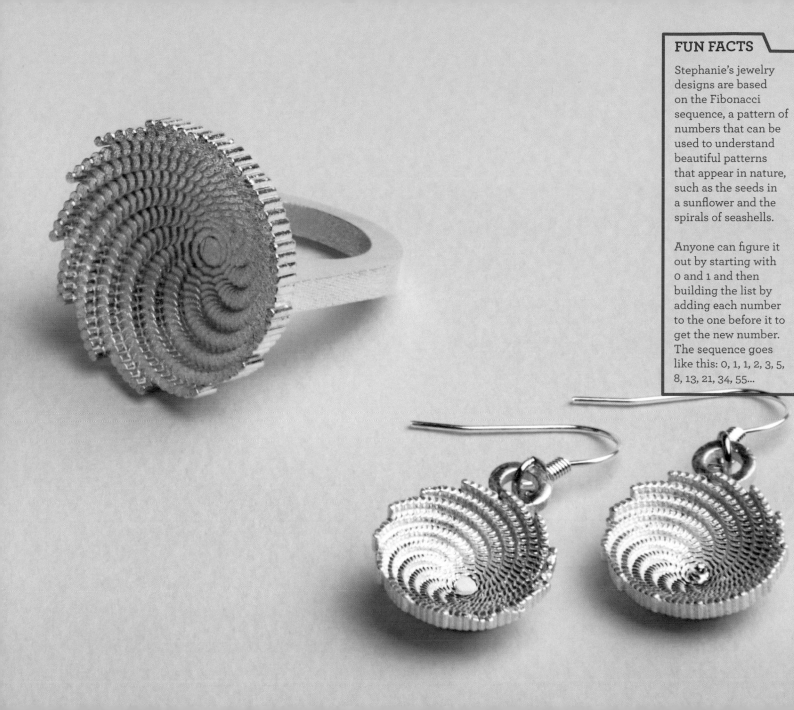

ALEXA
Age: 12
Location: New Utrecht
Likes: Jazz music
Talent: Tap dancing

SINCLAIR-10
[sĭn klâr tĕn]
Process: FDM
Description:
Internet-connected
printing appliance
Features:
Built-in screen,
Wi-Fi antenna

ALEXA &
SINCLAIR-10
THE PRODUCT-MAKING ROBOT

In the next neighborhood, to the south of the park, is Sinclair-10. He lives with a little girl named Alexa and her family. Sinclair-10 sits on the kitchen counter and displays a catalog of product designs to download from the Internet. Not only is he fun to play with, but what makes him really special is that he can help Alexa, the oldest of the three children, find clever and educational products when her family needs something new.

With Sinclair-10's help, Alexa looks through all the different kinds of products that people have created and uploaded as digital files. Then she can pick what she likes and have Sinclair-10 make them right away.

STUFF·A·VERSE

This morning, her mother told her she could get some toys to keep them occupied during the storm. Alexa decided to download musical instruments to create a band with her brothers. There's a shaker, a rubber band guitar and a special kind of flute called an ocarina. As soon as Sinclair-10 has finished building, all Alexa has to do is install the rubber bands and start composing.

"I took violin once, but was actually more fond of percussion." I seemed to be telling LEO all kinds of things about myself.

"All a little raucous for my tastes. But Sinclair-10 is quite suited to a higher volume on so many levels."

GEORGE &
AL1C3-D

THE STORE ROBOT

ALIC3-D [alĭs thrē dē]
Process: SLS
Description:
Humanoid with embedded
printing modules
Features:
Articulated hands,
programmed to handle
complaints with grace

GEORGE
Age: 25
Location: Downtown Brooklyn
Likes: fine wine and three-hour meals
Talent: Cultivating indoor gardens

I was beginning to feel wistful, wanting a 3D printer of my own, when LEO told me about George and AL1C3-D.

"AL1C3-D is so very lovely and works in an establishment in the busy downtown area. George just moved in to her neighborhood. He comes all the way from France!"

"France?!"

"France!"

"Brooklyn is so, so international."

"It certainly is. More and more every day. And even though George doesn't have his own robot, he loves to create his own products. So, just like printing paper at the copy store, at AL1C3-D's store, he can get physical objects printed from the designs he makes himself."

"Non?!"

"Mais, oui!"

"Just a few days ago, he wanted to add some personal touches to the kitchen in his apartment. He decided that the cabinet handles might be a good place to start. Instead of the plain knobs, he designed handles that also function as planters for edible plants called microgreens."

"Oh, I've had those before. They make a salad very fresh and flavorful."

"Well coming from France, George knows his food. And once he designed the planters on his computer, he uploaded them to the store, and the next time he's out for a walk, he can pick them up. AL1C3-D is very happy that George is in the neighborhood. She loves designers."

"Wow!" I had no idea there was so much going on in the world of robots. LEO was right. They were all over Brooklyn. And the world!

IRIS-7 [ĭrĭs sĕvĕn]
Process: SLA with up to
4 combined materials
Description: Combination
scanner-printer
Features: Wheeled mobility,
3D gesture recognition

NATHAN &
IRIS-7
THE CUSTOMIZER ROBOT

"We're really just getting started. The world is a very big place. As is Brooklyn. Take a look here. To the south, by the beach, is Nathan and IRIS-7. IRIS-7 is a robot that lets people make physical things that precisely suit their taste, style and size. She has built-in scanners that can measure a person's body and adjust a product so that when she makes it, the fit will be perfect. Gloves, headphones, and hearing aids; you name it and IRIS-7 can customize it. Nathan is one of her favorite collaborators."

"He is an adventurous young fellow who likes to take long walks and examine every grain of sand, shell and rock formation."

NATHAN
Age: 8
Location: Coney Island
Likes: Scary roller coasters
Talent: Can identify over 100 different sea creatures

"An explorer in the making, no doubt."

"Oh, yes. A very bright future indeed. He worked with IRIS-7 to make a special pair of sandals that perfectly fit the contours of his feet. Because they are made just for him, they are the most comfortable footwear he's ever worn."

"We all need a pair of those."

"No doubt! Although robots such as myself already come with built-in shoes. Plus, Nathan likes to leave his mark on the sand, so he can see where's he's been and his friends can find him when he's off exploring. He worked with IRIS-7 to create a sole for his sandals that leaves octopus prints behind wherever he walks."

"Nifty!"

"Nifty is right! His two best friends were so impressed that they asked IRIS-7 to make sandals for them. One pair to leave starfish prints, and the other to look like penguin feet."

I could tell LEO was very proud of how creative and useful his robot friends can be.

EMILIE &
CLAUDE-8
THE GOURMET ROBOT

CLAUDE-8 [kläd āt]
Process: FDM with pressurized food nozzles
Description: FDA-approved food service device
Features: Refrigerated storage compartment, 10 distinct flavor profiles

"All of this information is making me hungry. I could use a cupcake right about now."

"Well, if you're feeling hungry, you should meet my friend Claude-8, who lives near the bridge. He is a food making robot who can make things that are edible by squeezing food paste out of his nozzle."

"You're kidding, right?"

"No, we're not programmed to kid or jest, but we can tell jokes. Have you heard the one about the computer and the elephant?"

"That's okay. Maybe later. Tell me more about Claude-8."

Well, he likes to work with chefs like Emilie, who design fanciful foods that look and feel as exciting as they taste. Claude-8 lets her design incredible structures that can be printed in candy and chocolate. Even cheese.

"Candy, chocolate and cheese? I may just faint!"

EMILIE
Age: 35
Location: Williamsburg
Likes: Loves junk food
(as a guilty pleasure)
Talent: Makes the
most perfect soufflés

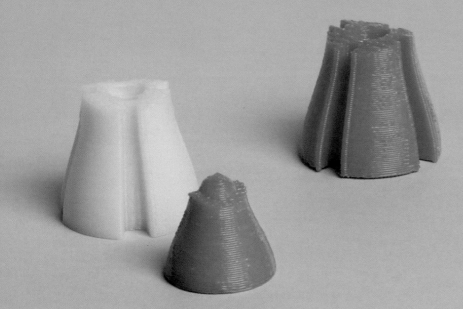

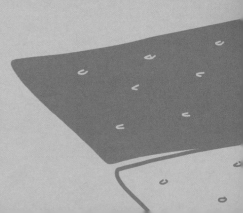

"Her latest creation is a special appetizer to get people to slow down and think while they nibble. She first lays out a chess board made of crackers, and then has Claude-8 build her a full chess set made of cheese. When a player captures a piece, their victory is exceptionally satisfying because they get to eat it!"

"They eat the cheese chess pieces?"
"They do. And oh, I wish I knew what that was like, but robots can't have cheese. It makes our gears seize up."

"No cheese? That is sad, LEO. Cheese is a true culinary treat."

"Sad, indeed. But tell me, Carla—a bit off topic, but I would really like to know—why is it that humans say the moon is made of cheese? It's not, you know. I have a robot friend named NiXie who lives on the moon, and has given me her firsthand report."

CHARLES & NIXIE

THE LUNAR-DUST ROBOT

CHARLES
Age: 6
Location: The moon
(previously Bedstuy)
Likes: Experimenting
with gravity
Talent: Hamster training

"You have a friend on the moon? That's so beyond Brooklyn, it's actually out of this world, you know."

"Yes, but not out of our solar system. We robots lead very exciting lives, especially NiXie, who is built for travel and adventure. NiXie lives with Charles and his mother, who is an astronautical architect. The whole family moved from directly in the center of Brooklyn to join her on the moon."

"That's almost too far to comprehend."

"238,900 miles to be exact. NiXie is very important to Charles' mom because she is a Lunar Dust Robot who can use what she finds on the surface of the moon to make new buildings. Since there is no electricity on the moon, she has solar panels to capture the sun's rays and convert it to the energy she needs to build. When NiXie is on a break from making moon buildings, she lets Charles play with her. He likes to make new kinds of houses for his hamster, Weenie."

NiXie [nikzē]
Process: Lunar dust SLS
Description:
Solar powered lunar dust sintering machine
Features:
GPS navigation, all-terrain chassis

"His latest design features a series of transparent capsules that can be stacked on top of one another, or connected end to end. One capsule has Weenie's sleeping chamber, and the other features a seesaw ladder so she can get her daily exercise. Charles is a very thorough and thoughtful hamster owner."

"And Weenie is one lucky hamster to have such a smart-looking and functional new home."

I could have kept listening forever to LEO tell me stories of all the other robots he knew and how they were working with people to make things to use in their everyday lives, but then it dawned on me...

"WAIT!" I said, holding up my hand. "If you can make me anything I can draw, then you can create the part I need to fix my bicycle chain!"

"Technically, I'm calibrated for art, but from what I understand, art and science and math and engineering are starting to become indistinguishable anyway. So, yes, I believe I can print you any part you need."

"YES!" I jumped up and down, and then hunkered close to the bike to figure out the shape of the chain link.

I grabbed my ball point once again and pulled out a fresh piece of paper. I held my hand as steady as I could and drew the link and the chain tool. Then I gave it to LEO and he scanned it like he had done with the sheep.

The tray went back up above his head, and he got busy. A few moments later, he presented the tray to me with the finished part.

I turned the bike upside down, repaired my chain and then tested the gear mechanism by turning the pedal.

Eureka! It worked! I had a working bicycle! I was saved and now I could get home.

But what about poor LEO, the 3D Prints, my new but misplaced royal robot friend? What would become of him, stranded out here in Brooklyn as the fierce storm was fast approaching?

And then I knew exactly what to do! It was a moment beyond Voila! and Eureka! I drew a cart with wheels to attach to my bike, and had LEO make it for me.

LEO looked a little confused.

"LEO, the weather's too rough to depend on a magic sheep. Hop in and I'll get you home."

"Oh, Carla, I am deeply touched. You are a very considerate human being. And a great creator."

As we bravely headed into the storm, I thought of my sculptures packed away deep in storage. Perhaps it was time to take the box off the shelf. Because a dream that's stored, can never be fully explored.

My friendship with LEO the Maker Prints continues to flourish well after recovery from the storm.

I draw, he prints, and together we make.

Visit **LeoTheMakerPrince.com** to learn more,
or go to **Thingiverse.com/LeoTheMakerPrince**
to download the objects in this book.

Special thanks to everyone who helped make this project happen:

My core team of dedicated and creative forces:
Alexa, Cindy, Claudia, and Nicholas

Mike Glaser, for his loving support

Brian Jepson and Dale Dougherty at Maker Media, SVA Visible Futures Lab, SVA Products of Design Program, Drexel University Product Design Program, Smart Design, Cornell University Creative Machines Lab, Alan Chochinov, Leif Mangelsen, Marko Manriquez, John T. Heida, Jeffrey Lipton, John Dimatos, Gisella Diana...

and Roo, the coolest dog in the world

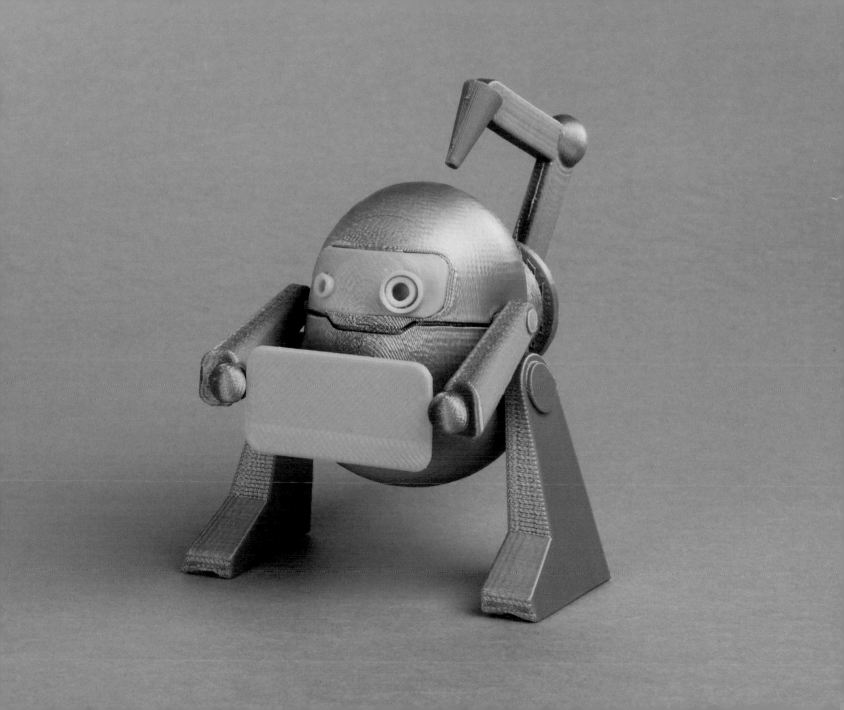